Anne Cazor & Christine Liénard

Photographs by Julien Attard
Translation by Gui Alinat

My
Molecular Cuisine
Kit

28 Recipes

CRC Press
Taylor & Francis Group
an **informa** business

www.crcpress.com

Introduction

Science explores the world, and researches the mechanisms of natural phenomena. Molecular Gastronomy is a scientific discipline studying culinary transformations, as well as phenomena already associated with gastronomy in general. Molecular Gastronomy is part of food science. Technology uses scientific knowledge to produce applications and, in the case of the culinary arts, new applications in the kitchen. Cooking, a mix of art and technique, is inseparable from taste, ingredient quality and craftsmanship. If, in addition, we seek to understand, then cooking becomes molecular cooking. Molecular cooking, or gastronomy, uses applications inspired from technology to create new dishes, new textures, new flavors, and new sensations...

Table of Contents

Emulsion

Puffed peanut chicken fries with Pastis mayonnaise	12	
Tuna club sandwich, sesame-vermouth white mayonnaise	14	
Strawberry-pesto-cider gazpacho with strawberry coulis	16	
Trio of vinaigrettes in spray bottles	18	

Soft gel

Apple and beet tea	20	
Frosty mint and chocolate marshmallow	22	
Sabayon mousse and cold ratatouille	24	

Brittle gel

Dulce de leche and crystal salt	26	
Baked camembert and honey pearls	28	
Foie gras, Muscat and dark chocolate lollipops	30	
Apple and cider vinegar crumble	32	
Gelled piña colada	34	

Elastic gel

Pulled duck confit with white wine and orange spaghetti	36	
Coco flan and Curaçao spaghetti	38	

Table of Contents

Elastic gel

Chocolate-balsamic macaroon — 40

Salmon sashimi and avocado, banana and balsamic gelée — 42

Spherification

Lychee sphere with ricotta, raspberry and rosewater crème — 44

Vodka shot with apple-caramel sphere — 46

Raw oyster and its raspberry vinegar pearl — 48

Deconstructed tiramisu — 50

Spherical tzatziki — 52

Cardamom and yogurt sphere, mango tartare — 54

Spherical chorizo and cider — 56

Kahlua® sphere and vodka shot — 58

Mousse

Crumble of raspberry and lychee mousse — 60

Frozen Nutella® mousse — 62

Boudin Créole crème with foie gras foam — 64

Avocado mousse on a bed of sweet polenta — 66

Ingredients

As you are about to discover in this book, you will be using texture agents such as agar-agar, sodium alginate, etc... These texture agents are used for their specific characteristics: to obtain a heat-resistant gel, for instance, which can't be done with gelatin, but can with agar-agar.

The specific ingredients offered in this book are food additives. A food additive is a substance with or without nutritional value, intentionally added to an ingredient with a specific goal in mind, whether that goal be technological, sanitary, organoleptic or nutritional.

There is no legal, quantitative limit to the consumption of food additives, except for carrageenan (maximum daily dose of 75 mg/kg of body mass). The food additives found in this book should only be used in the proportions described.

The consumption of food additives should be avoided for children under 6 years of age.

Agar-agar

A gelling agent extracted from red algae.
Widely used in Asian cuisine and produces heat-resistant hard gels (up to 80°C/176°F).
Recommended dosage: 1% (1 g per 100 g)

Sodium alginate

A thickening/gelling agent extracted from brown algae.
Increases viscosity of low-calcium mixtures and is used, in presence of calcium, for spherification purposes.
Note: Do not pour sodium alginate solutions down the drain as an obstruction is possible (sodium alginate may gel in contact with calcium contained in tap water).
Recommended dosage: 1% (1 g per 100 g)

Carrageenan

A gelling agent extracted from red algae.
Obtains heat-resistant, soft, elastic gels (up to 65°C/149°F). Safe daily dose for human consumption is limited to 75 mg per kg (34 mg per lb) of body weight.
Recommended dosage: 1% (1 g per 100 g)

Gelatin

A gelling agent extracted from animal bones. Obtains soft gels that melt when heated.
Recommended dosage: 3% (3 g per 100 g)

Calcium salt

A source of calcium (lactate and/or calcium gluconolactate).
Increases calcium level as needed for spherification purposes.
Recommended dosage: 1% (1 g per 100 g)

Techniques

Emulsion

An emulsion is a dispersion of two liquids that otherwise do not mix. The dispersed liquid forms droplets in the other. In Cooking, most well-known emulsions are oil and water emulsions. These are not stable emulsions since both liquids end up eventually separating.

Nonetheless, tensioactive molecules (soy lecithin, phospholipids, some proteins, such as gelatin, etc...) allow emulsions to stabilize when they position between the dispersed liquid droplets and each other, thus preventing both liquids from separating.

Soft gel

A gel is a liquid contained in a network. That network can be made up of proteins (such as gelatin, egg proteins, etc...) or polysaccharides (agar-agar, carrageenan, etc...). Gelatin is a protein extracted from meat or fish. This protein is a gelling agent and allows a change of texture from liquid to a gel, thanks to a network forming between the proteins of gelatin. Gelatin dissolves in hot preparations (above 50°C/122°F), and gels at temperatures around 10°C/50°F. If the gel is heated back up to 37°C/98.6°F, it will melt.

If air is added into a preparation while gelling, air bubbles then stabilize (gelatin positions between water and air bubbles) and the preparation gels as it cools. The new gel blocks air bubbles, thanks to the protein network, and we obtain a gelled mousse.

Brittle gel

Agar-agar is a gelling agent extracted from red algae. Unlike gelatin (a protein), this gelling agent is a polysaccharide (a sugar molecule). It dissolves in hot preparations containing water: Boiling 1 to 3 minutes is appropriate. It gels at about 35°C/95°F. Agar-agar gels are hard and slightly opaque. If it is reheated above 80°C/176°F, it melts.

Elastic gel

Carrageenan is a gelling agent extracted from red algae. Unlike gelatin (a protein), this gelling agent is a polysaccharide (a sugar molecule). It dissolves in hot preparations containing water above 80°C/176°F; from a practical standpoint, it is appropriate to bring the preparation to a boil. It gels at about 40°C/104°F. Carrageenan gels are elastic and transparent. If a carrageenan is reheated above 65°C/149°F, it melts.

Spherification

Spherification is the transforming of a liquid into spheres. This technique is achieved by using sodium alginate (a gelling agent extracted from brown algae), which has the property to gel when coming in contact with calcium. These gels are heat-resistant.

We distinguish two spherification techniques:
- Basic spherification: sodium alginate is solubilized in a preparation, then that preparation is submerged into a calcium bath. A film instantly gels around the preparation. We obtain a sphere with a liquid center, unstable in time (calcium progresses from the outside in and the sphere gels completely). These spheres must be consumed immediately.
- Reverse spherification: Here, calcium is naturally present in the preparation, and we submerge it into a bath of sodium alginate. A film instantly gels around the preparation. We obtain a sphere with a liquid center, stable in time once it is out of the sodium alginate bath.

Mousse

A mousse is a foamy emulsion. To obtain a mousse, an emulsion must contain a fat of thick consistency (at cold temperature). Incorporating gas in the preparation can be done by agitation or with a whipped cream dispenser. In both cases, the gas is confined in the emulsion because fat crystallizes when the temperature drops (either on ice or if the gas expands). We then obtain a mousse.

Recipes by Techniques

Emulsion

Soft gel

Brittle gel

Elastic gel

Spherification

Mousse

Puffed peanut chicken fries with Pastis mayonnaise

○○○ 6 servings ◑ 20 minute prep time ◔ 5 minute cook time

Ingredients

For the Pastis mayonnaise:
1 egg yolk
10 cl / ½ cup vegetable oil
1 tablespoon red wine vinegar
3 teaspoons Pastis
Salt and pepper to taste

For the chicken fries:
600 g / 21 oz boneless, skinless
  chicken breast
2 cups of puffed peanuts
2 eggs
200 g / 7 oz flour
4 tablespoons vegetable oil
Salt and pepper to taste

Preparation

Pastis mayonnaise:
○ Mix egg yolk and vinegar. Salt and pepper to taste.
○ Very slowly pour part of the oil to the egg mixture while whisking briskly, in order to emulsify the oil into the egg mixture. When the preparation emulsifies, whisk in the remaining oil.
○ Whisk in Pastis at the end. Season to taste.

Puffed peanut chicken fries:
○ Thinly grind puffed peanuts.
○ Add salt and pepper to taste.
○ Cut chicken breast into thick "fries."
○ Flour chicken fries, then add them to the eggs, previously beaten.
○ Bread them in the ground puffed peanuts.
○ Deep fry at 350°F for about 5 minutes.
○ Serve immediately with the Pastis mayonnaise on the side.

Explore

Fat (vegetable oil) is dispersed in water (contained in the egg yolk and vinegar) and forms tiny droplets. Tensioactive molecules (proteins contained in egg yolk) stabilize the emulsion when they position between fat and water droplets. The oil must first be added in small volumes to form an "oil in water" emulsion. If the volume of oil is incorporated too heavily, it will turn into a "water in oil" emulsion and the mayonnaise will not form.

Tuna club sandwich, sesame-vermouth white mayonnaise

000
000 6 servings ⏱ 10 minute prep time ◑ 30 minute rest time (optional)

Ingredients

Preparation

For the egg white mayonnaise:
1 egg white
7 cl / ⅓ cup vegetable oil
3 cl / 2 tablespoons sesame oil
2 cl / 1½ tablespoon Martini Rosso
  (vermouth)
Salt and pepper to taste

For the club sandwich:
200 g / 7 oz sashimi-grade tuna
10 cl / ½ cup Martini Rosso
  (vermouth) (optional)
12 slices of white bread
Toasted sesame seeds

For the egg white mayonnaise:
◦ Whisk together egg white and Martini Rosso. Salt and pepper to taste.
◦ Blend the vegetable and sesame oils together.
◦ Very slowly pour part of the resulting oil into the egg mixture while whisking briskly, in order to emulsify the oil into the egg mixture. When the preparation emulsifies, whisk in the remaining oil.
◦ Season to taste.
◦ Reserve.

For the club sandwich:
◦ Small dice sashimi-grade tuna.
◦ Marinate tuna in Martini Rosso for 30 minutes (optional).
◦ With a cookie cutter, cut the bread into 12 7-cm / 3-inch squares.

To assemble:
◦ Dry tuna if you marinated it (optional).
◦ Spread mayonnaise on bread.
◦ Place tuna on bread.
◦ Top with bread and add more tuna.
◦ Top with a third piece of bread.
◦ Sprinkle with sesame seeds and serve.

Explore

Without fat, tensioactive molecules (proteins contained in egg whites) stabilize the air added while whisking, which results in a mousse. If one incorporates fat (oil) to egg white proteins, these proteins will position between water (contained in the Martini Rosso and egg white) and fat, instead of positioning between water and air bubbles (as for making a mousse). The result is an aerated mayonnaise instead of a mousse.

Strawberry-pesto-cider gazpacho with strawberry coulis

000
000 6 servings　🕐 10 minute prep time　3 ❄ 3 hour cool time

Ingredients

For the gazpacho:
200 g / 7 oz strawberries
40 g / 1.5 oz pound cake
20 g / ¾ oz pine nuts
6 large leaves of fresh basil
5 cl / 4 tablespoons olive oil
2 cl / 1½ tablespoons cider
　vinegar
10 cl / ½ cup apple cider

For the strawberry coulis:
100 g / 3.5 oz strawberries
Sugar to taste

Preparation

Strawberry-pesto-cider gazpacho:
- In a stainless steel bowl, roughly mix strawberries which have been cut in half, pound cake, pine nuts, basil leaves, cider vinegar and olive oil. Cover and marinate in the refrigerator for at least 3 hours.
- Process with hand blender until smooth, then add apple cider to dilute.
- Finish mixing with hand blender until silky and well-emulsified.

Strawberry coulis:
- Process strawberries with hand blender until smooth.
- Add sugar to taste.

To assemble:
- Fill each shot glasses with gazpacho. Top with strawberry coulis.
- Serve chilled.

Explore

Fat (oil) is dispersed in water (in strawberries, vinegar, cider, etc...) to form tiny droplets. Phospholipids (tensioactive molecules in plant cell membranes) position between oil and water droplets and stabilize the emulsion. The volume of oil, in comparison to water's, is too weak to make a firm "oil in water" emulsion. The thick consistency of gazpacho is due to the added solid elements, such as strawberries, pound cake, etc...

Trio of vinaigrettes in spray bottles

◯◯◯ 6 servings 🕐 15 minute prep time 🕐 4 minute cook time

Ingredients

For the 1st vinaigrette:
2 cl / 1½ tablespoons balsamic
  vinegar
2 cl / 1½ tablespoons strawberry
  juice
2 cl / 1½ tablespoons vegetable
  oil
20 g / ¾ oz dark chocolate
5 g / 1½ teaspoon sugar

For the 2nd vinaigrette:
6 cl / 5 tablespoon orange juice
2 cl / 1½ tablespoon extra virgin
  olive oil
1 tablespoon honey
1 teaspoon Dijon mustard
Ground cumin to taste

For the 3rd vinaigrette:
5 cl / 4 tablespoon coconut milk
2 cl / 1½ tablespoon olive oil
20 g / ¾ oz white chocolate
5 g / 1½ teaspoon sugar
1 teaspoon wasabi powder
Green food dye

Preparation

For the 1st vinaigrette:
∘ Chop the dark chocolate and melt in a double boiler or in the microwave
  (2 minutes for 600-watt oven). Mix well until texture is smooth.
∘ Whisk in balsamic vinegar, then strawberry juice and sugar.
∘ Slowly pour in vegetable oil while whisking briskly.
∘ Serve to season a berry salad for instance.

For the 2nd vinaigrette:
∘ Mix Dijon mustard and orange juice together.
∘ Slowly pour in olive oil while whisking briskly. Add honey and cumin.
  Mix well.
∘ Serve to season crudités.

For the 3rd vinaigrette:
∘ Chop the white chocolate and melt in a double boiler or in the micro-
  wave (2 minutes for 600-watt oven). Mix well until texture is smooth.
∘ Whisk in coconut milk, then sugar and wasabi.
∘ Slowly pour in olive oil while whisking briskly. Add a few drops of food
  dye and mix well.
∘ Serve to season exotic fruit salad for instance.

Note: Chocolate fat crystallizes when cold. A vinaigrette containing chocolate should not be
stored in a refrigerator or else it will harden.

Explore

Dijon mustard and chocolate contain tensioactive molecules (phospholipids and soy lecithin). These molecules position between water (contained in orange juice, coconut milk, strawberry juice or balsamic vinegar) and fat (contained in oil, chocolate or coconut milk) and stabilize the emulsion.

Apple and beet tea

888 10 servings ◔ 15 minute prep time ◑ 30 minute cook time ◕ 20 minute rest time ²✳ 2 hour cool time

Ingredients

45 cl / 15 fl oz apple juice
15 cl / 5 fl oz water
1 cooked beet (about 125 g / 4½ oz)
30 to 50 g / 1 to 2 oz sugar (to taste)
2 Earl Grey tea bags
10 g / ⅓ oz gelatin (5 sheets)

Preparation

○ Soak gelatin sheets in cold water to soften.
○ In a sauce pan, simmer and reduce apple juice by ⅔ to obtain 15 cl / 5 fl oz of concentrated apple juice. Skim if necessary.
○ Bring water to a simmer and pour over cooked, diced beet. Let it infuse for 2 minutes and strain. Reserve 15 cl / 5 fl oz of beet juice.
○ In a sauce pan, heat beet juice and concentrated apple juice.
○ As soon as it simmers, take off the heat and infuse tea bags for 3 or 4 minutes. Remove tea bags.
○ Add sugar and reheat. As soon as it simmers again, take off the heat and immediately add the softened gelatin sheets. Whisk.
○ Pour the preparation into an adequate mold, in order to make a 2 to 3 cm (about 1 inch) layer. Let it cool at room temperature for 20 minutes, then in the refrigerator for 2 hours.
○ Slice the gel into cubes. Place 2 or 3 cubes in a cup or glass and add boiling water.
○ Mix and enjoy like a tea.

Explore

The preparation contains gelatin, which gels once chilled. Later on, the gel becomes liquid once boiling water is added and releases aromatics. If the "tea" cools down again, it will not gel, since the percentage of gelatin would be too weak to harness the added water.

Frosty mint and chocolate marshmallow

20 ● 20 pieces ◐ 30 minute prep time ◔ 8 minute cook time ● 1 hour rest time 2 ❄ 2 hour cooling time

Ingredients

300 g / 10 oz sugar + 1 teaspoon
20 cl / ¾ cup mint syrup
2 egg whites
16 g gelatin (8 sheets)
100 g / 4 oz powdered sugar
200 g / 7 oz dark chocolate

Preparation

◦ Soak gelatin sheets in cold water to soften.
◦ In a sauce pan, bring powdered sugar and mint syrup to a boil on moderate heat. Simmer for 7 to 8 minutes. Preparation will foam.
◦ Whip egg whites until soft peaks using an electric mixer. Add 1 teaspoon of sugar while whipping.
◦ Add gelatin to the mint syrup and whisk to dissolve gelatin.
◦ Slowly pour mint syrup into the egg whites and keep on mixing for 5 minutes.
◦ Pour a 2 to 3 cm (about 1 inch) layer of preparation into a rectangular mold.
◦ Cool at room temperature for 20 minutes, then chill in refrigerator for at least 2 hours.
◦ Unmold the marshmallow and dust with powdered sugar, so it's easier to handle.
◦ Dice the marshmallow into large square and dust each one with powdered sugar.
◦ Melt dark chocolate in a double boiler, on low heat. Stir from time to time. Once the chocolate is thoroughly melted, take off the heat and dip the dices in it (using little tongs or a toothpick), in order to coat them with a thin layer of chocolate.
◦ Reserve and let cool. Serve once the chocolate is set.

Explore

Beaten egg whites (composed of water and protein) result in a mousse. Adding gelatin to the mousse allows it to gel. We then obtain a gelled mousse.

The gelatin is previously dissolved in the syrup and not in the egg whites, which would present a risk of coagulation of egg white proteins during the heating process. The sweet syrup adds a sticky and elastic texture to the gelled mousse.

Sabayon mousse and cold ratatouille

⊙⊙ 4 servings ◑ 25 minute prep time ◑ 35 minute cook time ● 1 hour rest time 2⊛◑ 2.5 hour cool time

Ingredients

For the gelled mousse:
3 egg yolks
15 cl / 5 fl oz water
½ bouillon cube (chicken or
  vegetable)
6 g gelatin (3 sheets)

For the ratatouille:
1 green bell pepper
1 red bell pepper
4 tomatoes
1 large onion
2 garlic cloves
3 tablespoons extra virgin olive oil
Sugar to taste
1 teaspoon paprika
1 teaspoon ground coriander
Salt and pepper to taste

Preparation

Sabayon mousse:
- Soak gelatin sheets in cold water to soften.
- In a sauce pan, bring water and bouillon cube to a simmer.
- Add gelatin sheets and whisk.
- Reserve in a large stainless steel bowl and let cool at room temperature (about 30 minutes).
- Place the bowl on a cold water bath and slowly add the egg yolks while beating with an electric hand mixer, like a sabayon.
- Quickly transfer the preparation to individual ramekins and cool in the refrigerator for at least 2 hours.

Ratatouille:
- Small dice onions, tomatoes, red and green peppers. Mince garlic.
- Sauté onions and peppers in olive oil for 7 to 8 minutes on moderate heat.
- Add garlic and spices and cook 2 minutes.
- Add tomatoes, sugar and salt and cook partially covered for about 15 minutes, or until most of the water is evaporated. Season with salt and pepper.
- Let cool at room temperature, then chill.

To assemble:
- Unmold sabayon mousse and reserve.
- Spoon cold ratatouille into a clean ramekin. Top with sabayon mousse and serve.

Explore

Gelatin is dissolved in hot liquid, then egg yolks are added to the preparation. The gelatin is chosen for its tensioactive and gelling properties. On one hand, it allows for blending the liquid with the fat contained in the egg yolks, and, on the other hand, it gels the sabayon.

Dulce de leche and crystal salt

1 jar serving 10 minute prep time 30 minute cook time 1 hour rest time 30 minute cool time

Ingredients

For the crystal salt:
25 cl / 1 cup water
15 g / ½ oz salt
5 g / 2 teaspoons agar-agar

For the dulce de leche:
1 cup of water
1 7-oz can sweetened,
  condensed milk

Preparation

Crystal salt:
◦ In a small sauce pan, bring water and salt to a simmer. Sprinkle agar-agar and mix without incorporating too much air. Simmer 2 minutes. Stir from time to time.
◦ Take off the heat and pour in a mold.
◦ Let cool at room temperature for about 30 minutes, then chill in refrigerator for about 30 minutes.

Dulce de leche:
◦ In a cast iron pan, mix a cup of water with the contents of a 7-oz can of sweetened, condensed milk.
◦ Stir as you cook this mixture, and control the heat so it doesn't bubble over.
◦ Once it turns brown, turn the heat down a little until the mixture reaches your desired shade of color.
◦ Let it cool.

To serve:
◦ Spread dulce de leche on bread.
◦ Shave crystal salt on dulce de leche and enjoy.

Explore

In this recipe, it's the hard-gel properties of agar-agar that are displayed. An agar-agar brittle gel holds up well and is not elastic, which makes it easy to grate. It does not melt in the mouth and has a crunch, unlike a gelatin gel, for instance.

Agar-agar brittle gels are heat-resistant (up to 90°C/176°F) and can be served with hot dishes.

Baked camembert and honey pearls

88 4 servings ⏲ 15 minute prep time 🍵 35 minute cook time 🕐 10 minute rest time 2❄ 2 hour cooling time

Ingredients

For the honey pearls:
50 g / 3 tablespoons honey
5 cl / 3 tablespoons water
1 g / ⅓ teaspoon agar-agar
50 cl / 2 cups grape seed oil or
   very cold vegetable oil

For the baked camembert:
1 block of camembert

Preparation

Honey pearls:
◦ Place oil in refrigerator a few hours before the preparation.
◦ In a small sauce pan, heat water and honey. Sprinkle agar-agar and mix without incorporating too much air. Simmer 2 minutes. Stir from time to time.
◦ Take off heat and cool down at room temperature for 10 minutes.
◦ With a syringe or a pipette, take a sample of the preparation and plunge it drop-by-drop into the cold oil.
◦ Strain and rinse the honey pearls. Reserve.

Baked camembert:
◦ Preheat oven to 180°C/350°F.
◦ Wrap camembert in aluminum foil.
◦ Bake for 30 minutes.

To assemble:
◦ Take camembert out of the oven. Make a hole in the center of the cheese and add the honey pearls.
◦ Serve immediately.

Explore

Thanks to principles of physics, when an aqueous preparation is plunged in oil, it does not mix but naturally forms droplets. In this recipe, we explore fast-gelling capabilities of agar-agar. Using oil at a low temperature accelerates the cooling process and we obtain gelled droplets. Note that for very sweet preparations, it is necessary to dilute with enough water to allow the agar-agar to dissolve.

Foie gras, Muscat and dark chocolate lollipops

12 | 12 lollipops 20 minute prep time 5 minute cook time 10 minute rest time 30 minute cool time

Ingredients

150 g / 5 oz cooked foie-gras
50 g / 1¾ oz dark chocolate
10 cl / 3 tablespoons sweet white
  wine (Muscat, Riesling)
2 g / ¾ teaspoon agar-agar

Preparation

○ Grate or shave dark chocolate. Reserve.
○ With a melon baller, shape small (about 2-cm- / 1-inch-wide) balls of foie gras. Stick a toothpick or small wooden kebob in each ball.
○ In a small sauce pan, add the wine and bring to a simmer. Sprinkle agar-agar without incorporating too much air. Simmer 2 minutes. Stir from time to time.
○ Take off heat and cool at room temperature for about 10 minutes.
○ When preparation is thickening, quickly dip each lollipop into the wine gel, then again a second time. Roll it in the grated or shaved dark chocolate.
○ Reserve in refrigerator before serving.

Explore

In this recipe, we explore how agar-agar dissolves in alcohol. Wine contains alcohol, yet agar-agar still can form a network and gel. It is also possible to use agar-agar with acidic preparations, into which it can dissolve. Agar-agar is fast-gelling and is appropriate for coating.

Apple and cider vinegar crumble

Ingredients

For the crystal vinegar:
15 cl / 5 fl oz of cider vinegar
20 g / ¾ oz of brown sugar
2 teaspoons of vanilla extract
Ground nutmeg to taste
3 g / 1 teaspoon agar-agar

For the apples:
4 large apples
40 g / 1½ oz salted butter
40 g / 1½ oz sugar
12 oz sponge cake

Preparation

Crystal vinegar:
◦ In a small sauce pan, bring cider vinegar, brown sugar, vanilla and nutmeg to a simmer. Sprinkle agar-agar and whisk, without incorporating too much air. Simmer for 2 minutes. Stir from time to time.
◦ Take off heat and pour in a mold.
◦ Let cool at room temperature for 30 minutes, then refrigerate for 30 minutes.
◦ Grate or shave the vinegar hard gel.

Sautéed apples:
◦ Preheat oven to 180°C/350°F.
◦ Peel, core and small dice apples.
◦ In a sauté pan, heat salted butter and sugar on moderate heat. Add diced apples and cook 20 minutes. Stir from time to time.

To assemble:
◦ Place some sponge cake at the bottom of a glass.
◦ Spoon in sautéed apples and top with crystal vinegar.
◦ Enjoy.

Explore

In this recipe, we explore how agar-agar gels in acidic solutions. Vinegar is acidic (low pH), but agar-agar can still form a network and gel. This preparation creates a hard gel; once it's hard-gelled, it is possible to break it easily to mimic crystals.

Gelled piña colada

20 ● 20 candies ◷ 15 minute prep time ◷ 10 minute cook time ◷ 40 minute rest time ◷ 30 minute cool time

Ingredients

10 cl / ½ cup coconut milk
10 g / 1 tablespoon sugar
0.7 g / ¼ teaspoon agar-agar
8 cl / 3 fl oz white or dark rum
0.4 g / ⅛ teaspoon agar-agar
10 cl / ½ cup pineapple juice
0.7 g / ¼ teaspoon agar-agar
Brown sugar to taste

Preparation

◦ In a small sauce pan, bring coconut milk and sugar to a simmer on moderate heat. Sprinkle agar-agar and whisk, without incorporating too much air. Simmer for 2 minutes. Stir from time to time. Take off heat and pour preparation into a square or rectangular mold. Let cool at room temperature.
◦ 4 minutes after molding the above preparation, bring rum to a simmer on moderate heat. Add agar-agar the same as above. Simmer for 2 minutes. Stir from time to time. Take off heat. Cool preparation a few seconds and pour it on top of the coconut gel. Let cool at room temperature.
◦ 5 minutes after molding the rum preparation, bring pineapple juice to a simmer. Add agar-agar the same as above. Simmer 2 minutes. Stir from time to time. Take off heat. Let cool at room temperature for 3 minutes, then pour it on top of the rum gel. Once molded, let cool at room temperature for 30 minutes, then refrigerate 30 minutes.

To assemble:
◦ Unmold the gel using a knife if necessary. Cut into rectangles or squares.
◦ Sprinkle with brown sugar and serve immediately.

Explore

In this recipe, we take advantage of the properties of agar-agar to gel rapidly. Because it gels at 35°C/95°F, it is possible to assemble different layers. This recipe plays with the setting and melting temperatures of agar-agar. Precision is important: if preparation is too hot, layers will mix; if preparation is too cold, layers won't stick.

Pulled duck confit with white wine and orange spaghetti

888 6 servings ⏱ 15 minute prep time ⏱ 15 minute cook time

Ingredients

For the orange spaghetti:
15 cl / ½ cup orange juice
7.5 cl / 5 tablespoons white wine
  vinegar
75 g / 2½ oz sugar
4 g / 1 teaspoon carrageenan

For pulled duck confit:
1 leg of duck confit

Preparation

Orange spaghetti:
◦ In a saucepan, heat sugar and white vinegar (a process known as "gastrique") on moderate heat until it begins to caramelize (about 10 minutes).
◦ Add orange juice to the caramel (beware as the hot caramel may splatter).
◦ Reheat on low heat. Sprinkle carrageenan and whisk, without incorporating too much air.
◦ Connect a syringe (at least 20 ml) to a food-grade silicone tube (about 1 m / 3 ft).
◦ As the liquid starts simmering, take saucepan off the heat and plunge the other extremity of the syringe into the sauce, so you can suck it into the syringe.
◦ Once the tube is filled, plunge it in an ice bath for a few seconds while maintaining both ends of the syringe above the water line.
◦ Push the syringe and extract the orange spaghetti. Reserve.
◦ Repeat the process in order to make at least 2 other spaghetti servings. Reheat the sauce to liquefy it if needed.

Pulled duck confit:
◦ Take fat off and debone the duck confit.
◦ Pull the meat and reheat in the microwave.

To assemble:
◦ Plate orange spaghetti and top with pulled duck confit.

Note: Without a syringe or a tube, you can make similar tagliatelle if you pour the preparation onto a sheet pan and let cool at room temperature, then cut into ribbons.

Explore

In this recipe, we explore the elasticity of carrageenan gels. On one hand, carrageenan gels' elasticity is resistant to mistreatment. On the other hand, sugar increases elasticity. Therefore, handling this gel is fairly easy. In addition, carrageenan gels are heat resistant (up to 65°C/149°F) and can be served with hot dishes.

For sweet preparations (such as caramel), it becomes necessary to dilute them in enough water (orange juice, for instance) in order to make sure carrageenan is dissolved.

Coco flan and Curaçao spaghetti

000
000 6 servings 15 minute prep time 10 minute cook time 20 minute rest time 30 minute cool time

Ingredients

For the coco flan:
150 g / 5 oz sweetened
  condensed milk
30 cl / 10 fl oz coconut milk
30 g / 1 oz grated coconut
5 g / 1¼ teaspoon carrageenan

For the Curaçao spaghetti:
20 cl / ¾ cup Curaçao
4 g / 1 teaspoon carrageenan

Preparation

Coco flan:
○ In a sauce pan, bring coconut milk and sweetened condensed milk to a simmer. Sprinkle carrageenan while whisking, without incorporating too much air.
○ Add grated coconut, take off heat and pour into 6 small ramekins.
○ Let cool at room temperature (about 20 minutes), then refrigerate about 30 minutes.

Curaçao spaghetti:
○ In a small saucepan, heat up the Curaçao. Sprinkle carrageenan while whisking, without incorporating too much air.
○ Connect a syringe (at least 20 ml) to a food-grade silicone tube (about 1 m / 3 ft).
○ As it starts simmering, take saucepan off the heat and plunge the other extremity of the syringe into the Curaçao, so you can suck it into the syringe.
○ Once the tube is filled, plunge it in an ice bath for a few seconds, while maintaining both ends of the syringe above the water line.
○ Push the syringe and extract the Curaçao spaghetti. Reserve.
○ Repeat the process in order to make at least 2 other spaghetti servings. Reheat the sauce to liquefy it if needed.

To assemble:
○ Unmold the coco flans. Garnish them with Curaçao spaghetti and serve.

Note: Without a syringe or a tube, you can make similar tagliatelle if you pour the preparation onto a sheet pan and let cool at room temperature, then cut into ribbons.

Explore

In this recipe, we explore how carrageenan dissolves in alcohol. In spite of the alcohol content, the carrageenan still forms a network and gels the liquor. Elastic (spaghetti) and fast-gelling (flan) properties of carrageenan are also displayed. In addition, note that carrageenan is adequate to gel dairy-based preparations.

Chocolate-balsamic macaroon

6 6 macaroons 25 minute prep time 20 minute cook time 1 h 40 minute rest time

Ingredients

For the chocolate génoise:
3 egg whites + 4 egg yolks
40 g / 3 tablespoons sugar
  + 1 teaspoon
35 g / 1¼ oz brown sugar
75 g / 2⅕ oz flour
35 g / 1¼ oz melted butter
200 g / 7 oz dark chocolate

For the balsamic gelée:
25 cl / 1 cup balsamic vinegar
5 g / 1 teaspoon carrageenan

Preparation

Chocolate génoise:
◦ Preheat oven to 200°C/400°F.
◦ Layer a sheet pan with parchment paper and grease lightly with butter.
◦ Beat egg yolks with sugar and brown sugar for 5 minutes. Carefully fold in flour, then melted butter.
◦ With a mixer, beat egg whites to a soft peak and fold them into the egg yolks.
◦ Pour onto the sheet pan (1 cm / ½ inch thick). Bake 10 minutes.
◦ Turn génoise over on another sheet pan. Cover with a clean towel and let cool at room temperature for about 30 minutes.
◦ Melt dark chocolate in double boiler.
◦ Cut the génoise into 5 cm / 2 inch circles with a cookie cutter.
◦ Dip circles into the chocolate and let rest for 1 hour, or until chocolate has set.

Balsamic vinegar gelée:
◦ In a small saucepan, bring the vinegar to a simmer. Sprinkle carrageenan while whisking, without incorporating too much air.
◦ When it begins to simmer, take off the heat and pour into a mold (0.5 cm / ⅓ inch thick).
◦ Let cool at room temperature for about 20 minutes.

To assemble:
◦ Cut circles into the gelée with a cookie cutter (0.5 cm / ⅓ inch thick).
◦ Make macaroons with 2 circles of génoise and a circle of gélee in between. Serve.

Explore

In this recipe, we explore how carrageenan gels in acidic solutions. Vinegar is acidic (low pH), but carrageenan still forms a network and gels. It is also possible to use agar-agar in acidic preparations.

Salmon sashimi and avocado, banana and balsamic gelée

30 ● 30 candies ◐ 30 minute prep time ◓ 10 minute cook time ◑ 40 minute rest time

Ingredients

For the sweet soy sauce gelée:
5 cl / 3 tablespoons water
2.5 cl / 1½ tablespoons soy sauce
2.5 cl / 1½ tablespoons corn syrup
2 g / ½ teaspoon carrageenan

For the banana and balsamic gelée:
7 cl / ¼ cup banana juice
3 cl / ⅛ cup balsamic vinegar
2 g / ½ teaspoon carrageenan

For the assembling:
200 g / 7 oz skinless, boneless,
  sashimi-grade salmon
1 avocado

Preparation

Sweet soy sauce gelée:
- Mix soy sauce with corn syrup.
- In a small sauce pan bring water to a simmer. Sprinkle carrageenan while whisking, without incorporating too much air.
- At the first simmer, whisk in the soy sauce – corn syrup mixture. Take off the heat and pour a thin layer into a mold.
- Let cool at room temperature for about 20 minutes.

Banana-balsamic vinegar gelée:
- In a small sauce pan, bring to a simmer banana juice and balsamic vinegar. Sprinkle carrageenan while whisking, without incorporating too much air.
- At the first simmer, take off heat and pour a thin layer into a mold.
- Let cool at room temperature for about 20 minutes.

To assemble:
- Cut the salmon into 1-cm- / ⅓-inch-thick slices.
- Cut each slice into desired shape (square, round, etc...).
- Do the same with peeled, seeded avocado.
- Cut the 2 gelées in similar shapes, using the same cookie cutters.
- Assemble salmon and sweet soy sauce gelée.
- Assemble avocado and banana-balsamic vinegar gelée.
- Skewer with toothpicks and serve.

Explore

Carrageenan does not dissolve in soy sauce and corn syrup unless we proceed in two steps. First, it is dissolved in water. Then the sweetened soy sauce is progressively added next. This two-step technique works in preparations where carrageenan does not mix well.

Lychee sphere with ricotta, raspberry and rosewater crème

12 ● 12 spheres ◐ 15 minute prep time ◑ 30 minute rest time 2 ◉ 2 hour freeze time (optional)

Ingredients

For the ricotta, raspberry and rosewater crème:
100 g / 4 oz ricotta
50 g / 1¾ cups raspberry coulis
10 g / 1 tablespoon sugar
½ teaspoon rosewater

For the calcium bath:
30 cl / 10 fl oz water
3 g / 1 teaspoon calcium salt

For the lychee spheres:
10 cl / 6 tablespoons lychee juice
1 g / ¼ teaspoon sodium alginate

Preparation

Ricotta, raspberry and rosewater crème:
∘ Mix ricotta, raspberry coulis, sugar and rosewater together.
∘ Spoon adequate quantity in each spoon and refrigerate.

Calcium bath:
∘ Using an immersion blender, mix calcium salt into water until powder is completely dissolved. Rest 30 minutes.

Lychee spheres:
∘ Using an immersion blender, mix sodium alginate into lychee juice until powder is completely dissolved. Rest 30 minutes.
∘ Fill half-spherical silicone molds with the preparation and freeze for 2 hours. (This allows for homogenous spheres. This step is optional and a measuring spoon could be used with similar results.)
∘ Plunge the frozen half-spheres into the calcium bath for about a minute. (For better results, the spheres should not touch the sides of the container or float.)
∘ Then plunge them carefully into a plain water bath (use a small, slotted spoon to handle the spheres).

To assemble:
∘ Carefully place a sphere on each presentation spoon.
∘ Serve. Inform your guests to taste each preparation in one shot.

Note: Always dispose of sodium alginate in the trash, as opposed to into drains, which would cause an obstruction.

Explore

Sodium alginate is dissolved in lychee juice. Once the preparation is plunged into the calcium bath, a gelled film instantly forms around it and allows the spherification to occur. Even though spheres are rinsed, some calcium stays on the surface and will continue to gel from the outside in. At one point, the sphere will be entirely gelled.

Vodka shot with apple-caramel sphere

12 ● 12 spheres ◗ 20 minute prep time ◗ 30 minute rest time 2 ❄ 2 hour freeze time

Ingredients

For the calcium bath:
30 cl / 10 fl oz water
3 g / 1 teaspoon calcium salt

For the apple-caramel spheres:
20 cl / ¾ cup apple juice
5 cl / 3½ tablespoon caramel
  syrup
2.6 g / ½ teaspoon sodium
  alginate

To assemble:
Vodka

Preparation

Calcium bath:
◦ Using an immersion blender, mix calcium salt into water until powder is
  completely dissolved. Rest 30 minutes.

Apple-caramel spheres:
◦ Mix apple juice and caramel syrup.
◦ Using an immersion blender, mix sodium alginate into apple-caramel
  mix until powder is completely dissolved. Rest 30 minutes.
◦ Fill half-spherical silicone molds with the preparation and freeze for
  2 hours. (This allows for homogenous spheres. This step is optional and
  a measuring spoon could be used with similar results.)
◦ Plunge the frozen half-spheres into the calcium bath for about a minute.
  (For better results, the spheres should not touch the sides of the container
  or float.)
◦ Then plunge them carefully into a plain water bath (use a small, slotted
  spoon to handle the spheres).

To assemble:
◦ Fill half a shot glass with vodka.
◦ Carefully drop a sphere into each glass.
◦ Serve. Advise your guests to drink the vodka, then pop the sphere into
  their mouths.

Explore

In this recipe, sodium alginate is dissolved into the apple-caramel mix. When the half spheres are plunged into the calcium bath, a gelled film instantly forms around them and allows the spherification to occur. Even though spheres are rinsed, some calcium stays on the surface and will continue to gel from the outside in. At one point, the sphere will be entirely gelled.

Raw oyster and its raspberry vinegar pearl

888 6 servings ◔ 15 minute prep time ◑ 30 minute rest time

Ingredients

For the calcium bath:
30 cl / 10 fl oz water
3 g / 1 teaspoon calcium salt

For the vinegar pearls:
5 cl / 3 tablespoons raspberry
vinegar
10 cl / 7 tablespoons low-calcium
water (below 60 mg/l)
2 cl / 1⅓ tablespoons corn syrup
1.7 g / ½ teaspoon sodium
alginate
Red food coloring

To assemble:
3 dozen oysters

Preparation

Calcium bath:
◦ Using an immersion blender, mix calcium salt into water until powder is
  completely dissolved. Rest 30 minutes.

Raspberry vinegar pearls:
◦ Mix water and corn syrup.
◦ Using an immersion blender, mix sodium alginate into preparation until
  powder is completely dissolved.
◦ Add raspberry vinegar, then red food coloring. Rest 30 minutes.
◦ Just before serving, mix the preparation with a whisk.
◦ Take a sample with a syringe or a pipette and extract into the calcium
  bath drop by drop. After a few seconds, collect the pearls with a slotted
  spoon or small strainer, then plunge them carefully in a plain water bath.

To assemble:
◦ Place one or more vinegar pearls into oyster shell and serve immediately.

Note: Always dispose of sodium alginate in the trash, as opposed to into drains, which
would cause an obstruction.

Explore

In this recipe, the preparation is acidic (vinegar-based). Sodium alginate does not dissolve in acidic solutions, unless in two steps. Sodium alginate is dissolved in water first. Then vinegar is progressively added, which slowly reduces the volume of acid in the preparation containing sodium alginate. This two-step technique allows the spherification of acidic preparations.

Deconstructed tiramisu

888 6 servings ◐ 30 minute prep time ◐ 30 minute rest time 3 ✷ 3 hour cooling time

Ingredients

250 g / 9 oz mascarpone
40 g / ¼ cup sugar
2 large eggs
2 tablespoons orange marmalade
12 ladyfingers
Cocoa powder (optional)
Salt

For the calcium bath:
30 cl / 10 fl oz water
3 g / 1 teaspoon calcium salt

For the coffee pearls:
10 cl / 7 tablespoons strong coffee
(espresso)
2 cl / 4 teaspoons Amaretto
.6 g / ⅛ teaspoon sodium alginate

Preparation

Mascarpone cream:
◦ In a stainless steel bowl, whisk egg yolks with sugar until lighter in color.
◦ In another bowl, temper mascarpone with a wooden spoon until soft and smooth-textured. Add the egg-sugar mixture and orange marmalade. Mix and reserve.
◦ Beat egg whites with a pinch of salt to a soft peak.
◦ Whisk ¼ of egg whites into the mascarpone preparation. Then delicately fold in the rest of the whites in 2 batches.
◦ Place 2 ladyfingers at the bottom of each of 6 ramekins. Add a thick topping of mascarpone cream.
◦ Refrigerate for at least 3 hours.

Calcium bath:
◦ Using an immersion blender, mix calcium salt into water until powder is completely dissolved. Rest 30 minutes.

Coffee-Amaretto pearls:
◦ Sprinkle sodium alginate into hot coffee while whisking briskly until powder is completely dissolved (3 minutes).
◦ Whisk in Amaretto. Rest 30 minutes.
◦ Just before serving, take a sample with a syringe or a pipette and extract into the calcium bath drop by drop. After a few seconds, collect the pearls with a slotted spoon or small strainer, then plunge them carefully in a plain water bath.

To assemble:
◦ Dust ramekins with cocoa powder and top with coffee-Amaretto pearls.

Note: Always dispose of sodium alginate in the trash, as opposed to into drains, which would cause an obstruction.

Explore

The liquid that turns into spheres contains alcohol (Amaretto). Because sodium alginate does not dissolve in alcohol, let's explore other solutions:
- Dissolve sodium alginate in two steps: First in hot coffee, then Amaretto is progressively added (same method as acidic preparations).
- Evaporate alcohol: Amaretto is heated up with coffee; alcohol evaporates, which allows sodium alginate to dissolve.

Spherical tzatziki

Ingredients

For the sodium alginate:
30 cl / 1⅓ cup low-calcium water
 (below 60 mg/l)
1.5 g / ½ teaspoon sodium
 alginate

For the cucumber:
1 cucumber
1 teaspoon of olive oil
1 or 2 teaspoons lemon juice
Coarse sea salt

Greek yogurt spheres:
Greek yogurt
1 large garlic clove
Salt and pepper to taste

Preparation

Sodium alginate bath:
∘ Using an immersion blender, mix sodium alginate into water until powder is completely dissolved. Rest 30 minutes.

Cucumber:
∘ Peel and seed cucumber, then grate it or cut very thin slices. Sprinkle with salt and let cucumber release water and drain for at least 15 minutes. Rinse and dry.
∘ Add olive oil and lemon juice. Salt and pepper to taste.
∘ Plate grated or sliced cucumber in a porcelain spoon. Reserve.

Greek yogurt spheres:
∘ Mince garlic. Add to Greek yogurt. Salt and pepper to taste.
∘ Fill up the half-spherical silicone molds with Greek yogurt and freeze for at least 2 hours. (This allows for homogenous spheres. This step is optional and a measuring spoon could be used with similar results.)
∘ Plunge the frozen half-spheres into the sodium alginate bath for at least 30 seconds. (For better results, the spheres should not touch the sides of the container or float.)
∘ Then plunge them carefully in a plain water bath (use a small, slotted spoon to handle the spheres).

To assemble:
∘ Delicately place a sphere in each porcelain spoon and let it defrost if necessary.
∘ Serve. Inform your guests to taste each preparation in one bite.

Note: Always dispose of sodium alginate in the trash, as opposed to into drains, which would cause an obstruction.

Explore

Greek yogurt is naturally rich in calcium. When the yogurt preparation is submerged in sodium alginate, a film instantly gels around it and allows for spherification. Once the sphere is out of the sodium alginate bath, it stops gelling. Thanks to reversed spherification, the result is a sphere of Greek yogurt with a soft center.

Cardamom and yogurt sphere, mango tartare

12 ● 12 spheres 🕐 15 minute prep time 🌓 30 minute rest time 2 ❋ 2 hour freeze time (optional)

For the sodium alginate:
30 cl / 1⅓ cup low-calcium water
  (below 60 mg/l)
1.5 g / ½ teaspoon sodium
  alginate

For the mango tartare:
1 ripe mango

For the yogurt spheres:
Plain yogurt
5 g / 1 teaspoon sugar
1 pinch of ground cardamom

Sodium alginate bath:
◦ Using an immersion blender, mix sodium alginate into water until powder is completely dissolved. Rest 30 minutes.

Mango tartare:
◦ Peel mango and small dice.
◦ Plate in each porcelain spoon and refrigerate.

Yogurt and cardamom spheres:
◦ Mix yogurt, sugar and cardamom.
◦ Fill half-spherical silicone molds with yogurt and freeze for at least 2 hours. (This allows for homogenous spheres. This step is optional and a measuring spoon could be used with similar results.)
◦ Plunge the frozen half-spheres into the sodium alginate bath for at least 30 seconds. (For better results, the spheres should not touch the sides of the container or float.)
◦ Then plunge them carefully into a plain water bath (use a small, slotted spoon to handle the spheres).

To assemble:
◦ Delicately place one sphere on each porcelain spoon and let it defrost if necessary.
◦ Serve. Inform your guests to taste each preparation in one bite.

Note: Always dispose of sodium alginate in the trash, as opposed to into drains, which would cause an obstruction.

Explore

Yogurt is naturally rich in calcium. When the yogurt preparation is submerged in sodium alginate, a film instantly gels around it and allows for spherification. Once the sphere is out of the sodium alginate bath, it stops gelling. Thanks to reversed spherification, the result is a sphere of yogurt with a soft center.

Spherical chorizo and cider

12 ● 12 spheres 🕐 20 minute prep time 🕐 5 minute cook time 🕑 30 minute rest time 2 ❋ 2 hour freeze time

Ingredients

For the sodium alginate:
30 cl / 1⅓ cup low-calcium water
  (below 60 mg/l)
1.5 g / ½ teaspoon sodium
  alginate

For the chorizo spheres:
20 cl / ¾ cup whipping cream
60 g / 2 oz spicy chorizo

1 bottle of cider (alcohol)

Preparation

Sodium alginate bath:
○ Using an immersion blender, mix sodium alginate into water until powder is completely dissolved. Rest 30 minutes.

Spherical chorizo:
○ Small dice the chorizo.
○ In a small saucepan, heat cream and chorizo. As it begins to simmer, take off the heat and stir. Let cool at room temperature for about 20 minutes, then strain.
○ Fill up the half-spherical silicone molds with the liquid and freeze for 2 hours. (This allows for homogenous spheres. This step is optional and a measuring spoon could be used with similar results.)
○ Plunge the frozen half-spheres into the sodium alginate for about a minute. (For better results, the spheres should not touch the sides of the container or float.)
○ Then plunge them carefully into a plain water bath (use a small, slotted spoon to handle the spheres).
○ Carefully place a sphere on a spoon and let it defrost at room temperature if necessary.

To assemble:
○ Serve chilled cider in glasses with the spoons of chorizo-flavored cream on the side. Inform your guests to taste it in one bite.

Note: Always dispose of sodium alginate in the trash, as opposed to into drains, which would cause an obstruction.

Explore

Cream is naturally rich in calcium. When a cream-based preparation is plunged into a sodium alginate bath, a film gels around it and allows for spherification. Once the sphere is out of the sodium alginate bath, it will stop gelling. This technique of reversed spherification allows for a sphere of chorizo-flavored cream with a soft center.

Kahlua® sphere and vodka shot

12 ● 12 spheres   🕐 15 minute prep time   🌓 30 minute rest time   2 ❊ 2 hour freeze time

Ingredients

For the sodium alginate:
30 cl / 1⅓ cup low-calcium water
  (below 60 mg/l)
1.5 g / ½ teaspoon sodium
  alginate

For the Kahlua spheres:
60 g / 2 ½ cups whipping cream
4 cl / 3 tablespoons Kahlua

Vodka

Preparation

Sodium alginate bath:
◦ Using an immersion blender, mix sodium alginate into water until
  powder is completely dissolved. Rest 30 minutes.

Spherical chorizo:
◦ Whisk cream and Kahlua® together.
◦ Fill half-spherical silicone molds with the preparation and freeze for
  2 hours. (This allows for homogenous spheres. This step is optional and
  a measuring spoon could be used with similar results.)
◦ Plunge the frozen half-spheres into the sodium alginate for about a
  minute. (For better results, the spheres should not touch the sides of the
  container or float.)
◦ Then plunge them carefully into a plain water bath (use a small, slotted
  spoon to handle the spheres).

To assemble:
◦ Fill a shot glass half way with vodka. Carefully place a sphere in the glass
  and let it defrost at room temperature if necessary.
◦ Serve. Inform your guests to taste the preparation in one shot, and pop
  the sphere in their mouths.

Note: Always dispose of sodium alginate in the trash, as opposed to into drains, which
would cause an obstruction.

Explore

Cream is naturally rich in calcium. In this recipe, the preparation contains alcohol (Kahlua®). If too much alcohol is added, the gelled film might be too fragile to keep the sphere intact.

Crumble of raspberry and lychee mousse

░░░ 6 servings ◔ 10 minute prep time 2 ❄ 2 hour cooling time

Ingredients

For the raspberry mousse:
20 cl / ¾ cup cream
100 g / 3½ oz seedless raspberry
   coulis
20 g / 2 tablespoons sugar

For the crumble:
6 plain sugar cookies
18 lychees

Preparation

Raspberry mousse:
∘ Mix cream, raspberry coulis, and sugar.
∘ Pour the preparation into a whipped cream dispenser and refrigerate
   for 2 hours.

To assemble:
∘ Crumble a sugar cookie at the bottom of a glass. Add 3 diced lychees.
∘ Insert a gas cartridge into the cartridge node, shake the canister (upside
   down) and release mousse into the glasses.
∘ Serve immediately.

Explore

Cream is an emulsion containing water and fat, which is thick when cold. The emulsion is stable thanks to tensioactive agents (milk proteins). The raspberry coulis, which is mostly water, flavors the cream without altering its aeration. A whipped cream dispenser helps obtain a mousse. But the same result can be obtained by chilling and whisking the emulsion on ice.

Frozen Nutella® mousse

8888 6 servings ⏱ 10 minute prep time ⏱ 5 minute cook time 🌙 30 minute rest time 2❄ 2 hour cooling time 4❄ 4 hour freeze time

Ingredients

20 cl / ¾ cup whipping cream
100 g / 3½ oz Nutella + 60 g / 2 oz
6 mini ice cream cones (optional)
Candied almonds (optional)

Preparation

◦ In a sauce pan, heat cream and Nutella on moderate heat, and stir until Nutella® is entirely melted. Remove from heat.
◦ Let cook at room temperature (about 30 minutes)
◦ Blend with an immersion blender until preparation is smooth and silky. Pour it into a whipped cream dispenser and refrigerate for 2 hours.
◦ Insert a gas cartridge into the cartridge node, shake the canister (upside down) and proceed in one of two ways:

To make scoops:
◦ Pour the mousse in a container.
◦ Reheat leftover Nutella and use a fork to "trickle" some over the mousse.
◦ Freeze for at least 4 hours.
◦ Serve a scoop in a cone or a small porcelain bowl.

To make lollipops:
◦ Use shot glasses or special molds for lollipops. Line with plastic film if necessary. Fill with mousse. It is also possible to use the leftover Nutella® to create a soft center in the middle.
◦ Freeze for 4 hours.
◦ Unmold.
◦ Sprinkle with cracked candied nuts.
◦ Serve immediately.

Explore

Cream and Nutella are both emulsions containing water and fat, which is thick when cold. The emulsion is stable thanks to tensioactive agents (milk proteins and soy lecithin in the Nutella).

Using a whipped cream dispenser helps obtaining a Nutella-flavored frozen mousse. But the same result can be obtained by chilling and whisking the emulsion on ice.

Boudin Créole crème with foie gras foam

88 4 servings   20 minute prep time   20 minute cook time   2 ⊛ 2 hour cooling time

Ingredients

For the foie gras foam:
100 g / 3.5 oz foie gras terrine
7 cl / ⅓ cup apple juice

For the boudin créole:
2 large boudin créole
  (125 g / 4.4 oz each)
3 shallots
3 teaspoons duck fat
10 cl / ½ cup apple juice
Salt and pepper to taste

Preparation

Foie gras foam:
∘ Temper foie gras terrine at room temperature so it is soft enough to spread.
∘ Mix with apple juice.
∘ Pass preparation through a sieve; it must be very smooth so it doesn't obstruct the whipped cream dispenser.
∘ Pour the preparation into the whipped cream dispenser and refrigerate for 2 hours.

Boudin créole crème:
∘ Mince shallots and salt them.
∘ In a sauté pan, sweat shallots in duck fat on low heat. Take off the heat when they are translucent (about 10 minutes).
∘ In the same pan, sauté boudin créole on each side and poke them with a fork.
∘ Take off the skin.
∘ Use an immersion blender to mix skinless boudin créole, shallots, and apple juice. Salt and pepper to taste.

To assemble:
∘ In each espresso cup, pour the crème.
∘ Add 1 or 2 cartridges into the whipped cream dispenser, heavily shake and top the crème with the foie gras foam.
∘ Serve immediately.

Explore

The foie gras and apple juice preparation is an emulsion containing water (apple juice) and fat (foie gras). The small quantity of tensioactive molecules does not allow for a stable emulsion. Using a whipped cream dispenser is recommended for this foam.

Avocado mousse on a bed of sweet polenta

6 servings ⏱ 10 minute prep time ⏱ 10 minute cook time ❄ 2 hour cooling time

Ingredients

For the avocado mousse:
2 ripe avocados, peeled and
  seeded
20 cl / ¾ cup apple juice
3 tablespoon of honey
Juice of 1 lemon
Some cashews

For the sweet polenta:
90 g / 3 oz polenta
50 cl / 2 cups milk
50 g / ¼ cup sugar

Preparation

Avocado mousse:
◦ Process avocado, apple juice, honey and lemon juice in a food processor.
◦ Pass the preparation through a sieve; it must be very smooth so it doesn't obstruct the whipped cream dispenser.
◦ Pour the preparation into the whipped cream dispenser and refrigerate for 2 hours.

Sweet polenta:
◦ In a saucepan, bring milk and sugar to a simmer. Sprinkle polenta over the hot milk while whisking.
◦ Slowly cook while stirring, per the cooking instructions on the package.
◦ Pour into 6 glasses or ramekins, molding layers of about 1 to 2 cm / ¾ inch thick.
◦ Refrigerate for at least 2 hours.

To assemble:
◦ Insert a gas cartridge into the cartridge node, shake the canister (upside down) and fill the glasses/remekins containing the polenta.
◦ Sprinkle with a few cracked cashews and serve.

Explore

The avocado-apple juice mix is an emulsion containing water (apple juice) and fat (avocado). The small quantity of tensioactive molecules does not allow for a stable emulsion. Using a whipped cream dispenser is recommended for this foam.

Weight/dosage conversion

Each spoon must be used with the scoop-and-sweep method (fill and flush with a knife).

This conversion table is only given for your information. Powders have different characteristics (the granulometry is different, for instance) depending on suppliers. Therefore, we recommend using a precision scale (0.1 g increments).

Make sure you wash and dry spoons between the ingredients.

Example of calculation:
◦ If 0.8% means 0.8 g of powder for 100 g of preparation, how much agar-agar should we use for 300 g of preparation?

   0.8 g of agar-agar corresponds to 100 g of preparation.

◦ If x g of agar-agar corresponds to 300 g of preparation, we obtain:

   $x/0.8 = 300/100$. So $x = (300/100) \times 0.8 = 2.4$ g.

◦ If we refer to the following table conversion, one can measure 2.4 g of agar-agar using:

   One 0.63 ml spoon + one 1.25 ml spoon + one 2.5 ml spoon.

Grams/volume conversion table

	0.63 ml spoon ⅛ teaspoon	1.25 ml spoon ¼ teaspoon	2.5 ml spoon ½ teaspoon
Agar-agar	0.4	0.6	1.4
Sodium alginate	0.6	0.8	1.9
Calcium salt	0.5	0.7	1.5
Carrageenan	0.6	0.9	2

Index of Recipes

A

Apple and beet tea 20

Apple and cider vinegar crumble 32

Avocado mousse on a bed of sweet polenta 66

B

Baked camembert and honey pearls 28

Boudin Creole crème with foie gras foam 64

C

Cardamom and yogurt sphere, mango tartare 54

Chocolate-balsamic macaron 40

Coco flan and Curacao spaghetti 38

Crumble of raspberry and lychee mousse 60

D

Deconstructed tiramisu 50

Dulce de leche and crystal salt 26

F

Foie gras, muscat and dark chocolate lollipops 30

Frosty mint and chocolate marshmallow 22

Frozen Nutella® mousse 62

G

Gelled piña colada 34

K

Kahlua® sphere and vodka shot   58

L

Lychee sphere with ricotta, raspberry and rosewater crème   44

P

Puffed peanut chicken fries with Pastis mayonnaise   12
Pulled duck confit with white wine and orange spaghetti   36

R

Raw oyster and its raspberry vinegar pearl   48

S

Sabayon mousse and cold ratatouille   24
Salmon sashimi and avocado, banana and balsamic gelée   42
Spherical chorizo and cider   56
Spherical tzatziki   52
Strawberry-pesto-cider gazpacho with strawberry coulis   16

T

Trio of vinaigrettes in spray bottles   18
Tuna club sandwich, sesame-vermouth white mayonnaise   14

V

Vodka shot with apple-caramel sphere   46

CRC Press
Taylor & Francis Group
6000 Broken Sound Parkway NW, Suite 300
Boca Raton, FL 33487-2742

International Standard Book Number: 978-1-4398-7942-9 (Hardback)

**Visit the Taylor & Francis Web site at
http://www.taylorandfrancis.com**

**and the CRC Press Web site at
http://www.crcpress.com**